CENGAGE Learning™

Chemistry Laboratory Notebook

This laboratory notebook belongs to

Name _____

Address _____

Telephone number _____

Course _____ Section _____

Laboratory instructor's name _____

Laboratory partner's name _____

Equipment drawer or locker number _____

Date: from_____ to _____

T0156182

Telephone Numbers of Emergency Services

Campus police _____

Campus fire _____

Health center _____

Poison control center ___   _____

Suggestions for Using This Laboratory Notebook

1. Fill in your name, address, and telephone number on the first page so your notebook can be returned if lost.

2. Create a table of contents for your notebook as you use it during the course. Record the titles of the experiments next to the page numbers on which your write-up can be found.

3. Notebook entries are a **permanent record** of your laboratory work. Never remove the original (white) pages. The yellow (copy) pages are perforated for easy removal, if you are required to turn in a copy of your work.

4. Always **write in ink** using a firm-tipped pen, such as a ballpoint. Use enough pressure to ensure that a clear, dark copy is made.

5. If you make a mistake, draw a single line through the error and write the correct entry nearby.

6. Provide the information asked for at the top of each notebook page.

7. Use the left column to record data and calculations. In the right column write a report of your work. Include all of the elements required by your laboratory instructor, such as the purpose of the experiment, the procedure used, observations made, and conclusions drawn.

Table of Contents

26 _____

27 _____

28 _____

29 _____

30 _____

31 _____

32 _____

33 _____

34 _____

35 _____

36 _____

37 _____

38 _____

39 _____

40 _____

41 _____

42 _____

43 _____

44 _____

45 _____

46 _____

47 _____

48 _____

49 _____

50 _____

Safety Contract*

Whenever I am in an area where laboratory reagents are being used, I agree to abide by the following rules:

1. Wear safety goggles.

2. Wear proper clothing.

3. Use good housekeeping practices.

4. Do only authorized experiments, and work only when the laboratory instructor or another qualified person is present.

5. Treat laboratory reagents as if they are poisonous and corrosive.

6. Dispense reagents carefully. Dispose of laboratory reagents as directed.

7. Not eat, drink, use tobacco, or apply cosmetics in the laboratory.

8. Report all incidents to the laboratory instructor.

9. Be familiar with the location and use of all safety equipment.

10. Become familiar with each laboratory assignment before coming to the laboratory.

11. Anticipate the common hazards that may be encountered in laboratory.

12. Become familiar with actions to be taken in the event of incidents in the laboratory.

student signature _____ date _____

laboratory instructor _____ date _____

In the space below, give any health information, such as pregnancy or other circumstance, that might help the laboratory instructor provide a safer environment for you, or that could aid the laboratory instructor in responding to an incident involving you in the laboratory.

1. I do/do not (circle one) expect to wear contact lenses during laboratory work. [Note: Goggles must still be worn when contact lenses are worn.]

2. List any known allergies to medication or other chemicals.

*From Rapp, M.W. *Practicing Safety in the Organic Chemistry Laboratory*; Chemical Education Resources: Palmyra, PA, 1997

Experiment title and number					Date		1
Name			Course	Section	Lab partner		

Experiment title and number				Date		1
Name			Course	Section	Lab partner	

Experiment title and number		Date
Name	Course	Section	Lab partner

Experiment title and number		Date		2
Name		Course	Section	Lab partner

Experiment title and number

Date

Name

Course

Section

Lab partner

Experiment title and number

Date

3

Name

Course

Section

Lab partner

Experiment title and number		Date	
Name	Course	Section	Lab partner

Experiment title and number		Date	
Name		Course	Section	Lab partner

4

Experiment title and number		Date	5
Name	Course	Section	Lab partner

Name | Course | Section | Lab partner

Experiment title and number

Date

6

Name

Course

Section

Lab partner

6

Experiment title and number		Date
Name	Course	Section	Lab partner

Experiment title and number

Date

Name

Course

Section

Lab partner

Experiment title and number

Date

Name

Course

Section

Lab partner

Experiment title and number

Date

Name

Course

Section

Lab partner

Caution: Place fold-in flap under yellow sheet before writing, to protect the pages that follow.

Experiment title and number

Date

8

Name

Course

Section

Lab partner

Experiment title and number

Date

Name

Course

Section

Lab partner

Caution: Place fold-in flap under yellow sheet before writing, to protect the pages that follow.

Experiment title and number

Date

9

Name

Course

Section

Lab partner

Caution: Place fold-in flap under yellow sheet before writing, to protect the pages that follow.

Experiment title and number

Date

Name

Course

Section

Lab partner

Experiment title and number		Date
Name	Course	Section	Lab partner

Experiment title and number

Date

11

Name

Course

Section

Lab partner

Experiment title and number

Date

11

Name

Course

Section

Lab partner

Caution: Place fold-in flap under yellow sheet before writing, to protect the pages that follow.

Experiment title and number		Date	
Name		Course	Section	Lab partner

Experiment title and number		Date
Name	Course	Section	Lab partner

Experiment title and number

Date

13

Name

Course

Section

Lab partner

Experiment title and number

Date

Name

Course

Section

Lab partner

Experiment title and number

Name

Caution: Place fold-in flap under yellow sheet before writing, to protect the pages that follow.

Experiment title and number		Date	
Name		Course	Section	Lab partner

Experiment title and number		Date	
Name		Course	Section	Lab partner

Experiment title and number		Date
Name	Course	Section	Lab partner

Experiment title and number		Date	
Name		Course	Section	Lab partner

Experiment title and number

Date

16

Name

Course

Section

Lab partner

Caution: Place fold-in flap under yellow sheet before writing, to protect the pages that follow.

Experiment title and number		Date	16
Name	Course	Section	Lab partner

| Experiment title and number | | | | | | | | Date | | | | | | | | 17 |

| Name | | | | | | Course | | Section | Lab partner | | | | | |

Experiment title and number		Date
Name	Course	Section	Lab partner

Experiment title and number		Date	
Name		Course	Section	Lab partner

Caution: Place fold-in flap under yellow sheet before writing, to protect the pages that follow.

Experiment title and number

Date

Name

Course

Section

Lab partner

Experiment title and number

Date

Name

Course

Section

Lab partner

Caution: Place fold-in flap under yellow sheet before writing, to protect the pages that follow.

Experiment title and number

Date

Name

Course

Section

Lab partner

Experiment title and number		Date

Name   Course   Section   Lab partner

Experiment title and number

Date

Name

Course

Section

Lab partner

Experiment title and number		Date
Name	Course	Section	Lab partner

Caution: Place fold-in flap under yellow sheet before writing, to protect the pages that follow.

Experiment title and number

Date

Name

Course

Section

Lab partner

Experiment title and number		Date	
Name	Course	Section	Lab partner

Experiment title and number		Date

Name		Course	Section	Lab partner

Experiment title and number

Date

23

Name

Course

Section

Lab partner

Caution: Place fold-in flap under yellow sheet before writing, to protect the pages that follow.

Experiment title and number		Date	
Name	Course	Section	Lab partner

Experiment title and number

Date

Name

Course

Section

Lab partner

Caution: Place fold-in flap under yellow sheet before writing, to protect the pages that follow.

24

Experiment title and number		Date	
Name	Course	Section	Lab partner

Caution: Place fold-in flap under yellow sheet before writing, to protect the pages that follow.

Experiment title and number

Date

25

Name

Course

Section

Lab partner

Experiment title and number		Date
Name	Course	Section	Lab partner

Experiment title and number		Date	
Name	Course	Section	Lab partner

Experiment title and number		Date	
Name	Course	Section	Lab partner

Experiment title and number		Date

Name	Course	Section	Lab partner

Experiment title and number

Date

Name

Course

Section

Lab partner

Experiment title and number

Date

Name

Course

Section

Lab partner

Caution: Place fold-in flap under yellow sheet before writing, to protect the pages that follow.

Experiment title and number

Date

Name

Course

Section

Lab partner

Experiment title and number

Date

Name

Course

Section

Lab partner

Experiment title and number		Date	
Name	Course	Section	Lab partner

Experiment title and number

Date

Name

Course

Section

Lab partner

Caution: Place fold-in flap under yellow sheet before writing, to protect the pages that follow.

0

Experiment title and number   Date

Name   Course   Section   Lab partner

Experiment title and number

Date

Name

Course

Section

Lab partner

Caution: Place fold-in flap under yellow sheet before writing, to protect the pages that follow.

Experiment title and number		Date
Name	Course	Section	Lab partner

Caution: Place fold-in flap under yellow sheet before writing, to protect the pages that follow.

Experiment title and number		Date

Name	Course	Section	Lab partner

Experiment title and number

Date

Name

Course

Section

Lab partner

Experiment title and number

Date

Name

Course

Section

Lab partner

Caution: Place fold-in flap under yellow sheet before writing, to protect the pages that follow.

Experiment title and number

Date

Name

Course

Section

Lab partner

Caution: Place fold-in flap under yellow sheet before writing, to protect the pages that follow.

Name

Course

Section

Lab partner

Caution: Place fold-in flap under yellow sheet before writing, to protect the pages that follow.

Experiment title and number		Date	
Name	Course	Section	Lab partner

Experiment title and number

Date

Name

Course

Section

Lab partner

Experiment title and number		Date

Name	Course	Section	Lab partner

Caution: Place fold-in flap under yellow sheet before writing, to protect the pages that follow.

Experiment title and number		Date	
Name		Course	Section	Lab partner

Experiment title and number		Date	36
Name	Course	Section	Lab partner

Experiment title and number

Date

Name

Course

Section

Lab partner

Experiment title and number		Date	
Name		Course	Section	Lab partner

Experiment title and number		Date
Name	Course	Section	Lab partner

Experiment title and number

Date

Name

Course

Section

Lab partner

Experiment title and number	Date
Name	Course	Section	Lab partner

Experiment title and number

Date

Name

Course

Section

Lab partner

Experiment title and number		Date
Name	Course	Section	Lab partner

Caution: Place fold-in flap under yellow sheet before writing, to protect the pages that follow.

Experiment title and number		Date
Name	Course	Section	Lab partner

Caution: Place fold-in flap under yellow sheet before writing, to protect the pages that follow.

Experiment title and number

Date

41

Name

Course

Section

Lab partner

Experiment title and number

Date

Name

Course

Section

Lab partner

Experiment title and number		Date

Name	Course	Section	Lab partner

Caution: Place fold-in flap under yellow sheet before writing, to protect the pages that follow.

Experiment title and number		Date	
Name	Course	Section	Lab partner

Experiment title and number

Date

Name

Course

Section

Lab partner

Experiment title and number

Date

Name

Course

Section

Lab partner

Caution: Place fold-in flap under yellow sheet before writing, to protect the pages that follow.

Experiment title and number

Date

44

Name

Course

Section

Lab partner

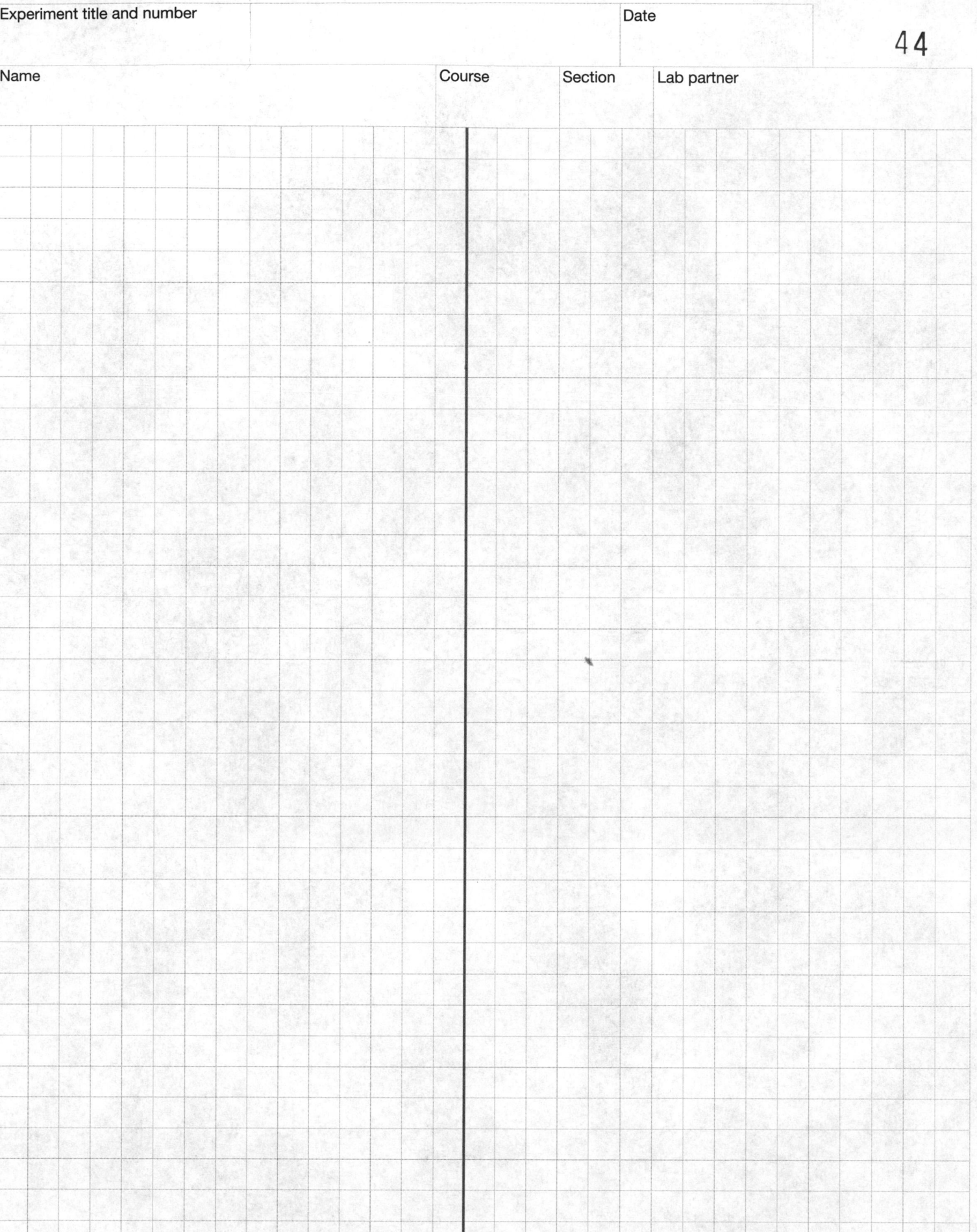

Experiment title and number		Date	44
Name	Course	Section	Lab partner

Experiment title and number		Date

Name	Course	Section	Lab partner

Experiment title and number		Date	
Name	Course	Section	Lab partner

Experiment title and number		Date

Name		Course	Section	Lab partner

Experiment title and number		Date
Name	Course	Section	Lab partner

Caution: Place fold-in flap under yellow sheet before writing, to protect the pages that follow.

Experiment title and number		Date
Name	Course	Section	Lab partner

Caution: Place fold-in flap under yellow sheet before writing, to protect the pages that follow.

Experiment title and number		Date
Name	Course	Section	Lab partner

Caution: Place fold-in flap under yellow sheet before writing, to protect the pages that follow.

Experiment title and number

Date

Name

Course

Section

Lab partner

Experiment title and number

Date

Name

Course

Section

Lab partner

Experiment title and number

Date

Name

Course

Section

Lab partner

Experiment title and number		Date	
Name		Course	Section	Lab partner

Caution: Place fold-in flap under yellow sheet before writing, to protect the pages that follow.

Experiment title and number		Date
Name	Course	Section	Lab partner

Caution: Place fold-in flap under yellow sheet before writing, to protect the pages that follow.